# WATER CYCLE

## NATURE'S CYCLES

Ray James

Rourke
Publishing LLC
Vero Beach, Florida 32964

www.rourkepublishing.com

PHOTO CREDITS: All Photographs © Lynn M. Stone, except p. 5, 6, 13, 11 © Rourke Publishing

Editor: Robert Stengard-Olliges

Cover and interior design by Nicola Stratford

**Library of Congress Cataloging-in-Publication Data**

James, Ray.
  Water cycle / Ray James.
     p. cm. --  (Nature's cycle)
  ISBN 1-60044-182-3 (hardcover)
  ISBN 1-59515-534-1 (softcover)
  1.  Hydrologic cycle--Juvenile literature.  I. Title. II. Series: James, Ray. Nature's cycle.

GB848.S76 2007
551.48--dc22                                          2006014431

Printed in the USA

CG/CG

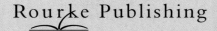

Rourke Publishing

www.rourkepublishing.com – sales@rourkepublishing.com
Post Office Box 3328, Vero Beach, FL 32964

# Table of Contents

# Water on Earth

Water seems to be almost everywhere. Water covers about seven of every ten parts of the Earth's surface.

Our bodies are made mostly of water. Without water, there would be no life on Earth.

The amount of water on Earth does not change. It is the same now as it has always been

However, like a top, water keeps going around. It moves from place to place as liquid, ice, or **vapor**.

Water vapor is a **gas**. It is invisible. It is part of the air around us.

# Evaporation

The story of water's journey is called the water cycle. The story begins with **evaporation**.

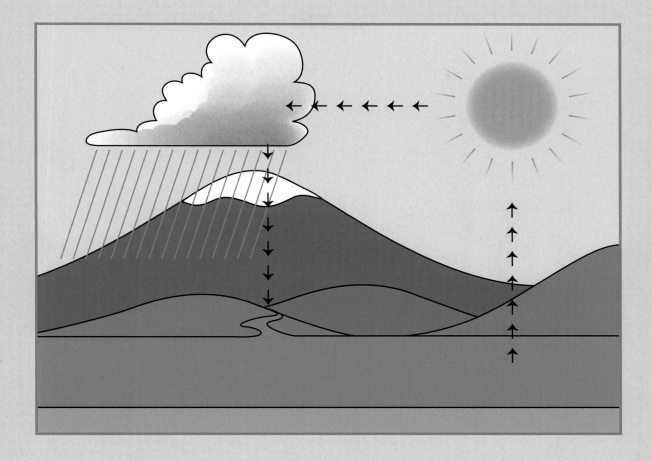

Evaporation turns liquid water into water vapor. Sunshine and heat speed up evaporation.

13

In its vapor form, water rises into the air. You cannot
see evaporation.

Water vapor is part of the air around us. It can change from its gas form.

# Condensation

Cold air cools water vapor. Cooling changes vapor back to liquid or ice.

This cooling process is called **condensation**. Droplets of water in the air form clouds.

Clouds can become heavy with water. Then they release it.

# Precipitation

Water may fall back to Earth as rain. It may fall as chips of ice or flakes of snow. These forms of falling water are called **precipitation**.

Precipitation falls on land and water. It does not fall in equal amounts everywhere.

# Glossary

**condensation** (kon den SAY shuhn) — water formed from a cooled gas

**evaporation** (e VAP u ray shuhn) — liquid changing into a gas or vapor

**gas** (GASS) — an invisible substance like air

**precipitation** (pri sip i TAY shuhn) — water falling from the sky as rain, sleet, hail, or snow

**vapor** (VAY pur) — mist, steam, or fog hanging in the air

# INDEX

## FURTHER READING

Kalman, Bobbie. *Water Cycle*.Crabtree, 2006.
Olien, Rebecca. *Water Cycle*. Capstone, 2005.

## WEBSITES TO VISIT

http://www.kidzone.ws/water/.html
http://www.epa.gov/OGWDW/kids_k_3.html

## ABOUT THE AUTHOR

Ray James writes children's fiction and nonfiction. A former teacher, Ray understands what kids like to read. Ray lives with his wife and three cats in Gary, Indiana.